Scott is an artist. He works with wood.

Scott learned how to make wooden animals.

Early on, Scott made a yellowish-brown duck.

Today Scott makes several kinds of birds.

He usually makes birds of America.

Scott had a top-notch show of wooden animals.

Many people heard of his show.

"I have the best job on earth!" Scott said.